Objective Questions in 'A' Level Chemistry

Answers

Objective Questions in 'A' Level Chemistry Answers

John Gunnell   Edgar Jenkins

*Centre for Studies in Science Education,
University of Leeds*

Oliver & Boyd

OLIVER AND BOYD
Tweeddale Court
14 High Street
Edinburgh EH1 1YL
A Division of Longman Group Limited

ISBN 0 05 002660 7

© John Gunnell and Edgar Jenkins, 1972

All Rights Reserved. No part of this publication may be reproduced, stored in a retrieval system, or transmitted, in any form or by any means, electronic, mechanical, photocopying, recording or otherwise, without the prior permission of the Copyright owners.

Filmset by Typesetting Services Ltd., Glasgow, Scotland

Printed by T. & A. Constable Ltd., Edinburgh

Contents

	Introduction	vii
	Summary of directions for answering the questions	ix
1	Atomic Structure	1
2	Structure and Bonding	2
3	The Gaseous State	3
4	Energetics	5
5	Kinetics	7
6	Equilibria I	8
7	Equilibria II	10
8	The Periodic Table	11
9	The Chemistry of the Non-Metals	12
10	The Chemistry of the Metals	13
11	General Inorganic Chemistry	14
12	Hydrocarbons	15
13	Compounds of Carbon, Hydrogen and Oxygen	16
14	The Organic Chemistry of Nitrogen and the Halogens	17
15	General Organic Chemistry	18
16	Experimental Procedures	19
17	Revision Paper I	20
18	Revision Paper II	21

Introduction

The decision to publish separately the answers to our *Objective Questions in 'A' Level Chemistry* reflects our belief that teachers and students use objective questions in a variety of ways.

The production of an answer book has enabled us to accommodate a suggestion received from several of those who helped in the pretesting of the materials. It was felt that immediate access to data or other information would have helped in the discussion of issues arising from the use of some of the questions in class. Accordingly, we have provided numerical data, worked solutions to calculations and added other pertinent comments.

It should be emphasised that all the information necessary to answer the questions is contained in the question papers themselves.

John Gunnell
Edgar Jenkins
Centre for Studies in Science Education,
University of Leeds

Summary of directions for answering questions

Multiple choice Only one response is correct. Indicate your choice of A, B, C, D or E on your answer sheet.

Multiple completion One or more of the responses given are correct. Indicate your choice of A, B, C, D or E as follows:

A	B	C	D	E
(*i*), (*ii*) and (*iii*)	(*i*) and (*iii*)	(*ii*) and (*iv*)	(*iv*)	(*i*) and (*iv*)

Matching pairs From the list A–E, choose the responses which correctly answer each of the questions in the group. Within each group, each of the letters A–E may be used once, more than once or not at all.

Assertion-reason questions Choose A, B, C, D or E as follows:

	Assertion	Reason	Argument
A	True	True	Reason is a *correct* explanation of assertion
B	True	True	Reason is *not a correct* explanation of assertion
C	True	False	Not applicable
D	False	True	Not applicable
E	False	False	Not applicable

1 Atomic Structure

1 C The nine (3^2) orbitals are $3s$, p_x, p_y, p_z, d_{xy}, d_{xz}, d_{yz}, $d_{x^2-y^2}$ and d_{z^2}.
2 B Be (ground state) $1s^2 2s^2$.
3 B 12 days is twice the half-life of the substance.
4 B 24 hours $= 3 \times (t_{\frac{1}{2}})P = 2 \times (t_{\frac{1}{2}})Q$.
      Considering 1 g of P and 1 g of Q,
      after 24 hours there remains $\frac{1}{8}$ g P and $\frac{1}{4}$ g Q.
5 C After $2 \times (t_{\frac{1}{2}})$ Rate (disintegrations $\min^{-1} g^{-1}$) $= \frac{15.3}{4} \simeq 3.8$ and

      after $3 \times (t_{\frac{1}{2}})$, $\frac{15.3}{8} \simeq 1.9$

      The given rate is between $2t_{\frac{1}{2}}$ and $3t_{\frac{1}{2}}$ but closer to the latter.
6 E
7 D 2.24 litres nitrogen gas at s.t.p.$= 0.1$ mole; each molecule contains 14 protons.
8 C Nitrogen (electron configuration $1s^2 2s^2 p_x^1 p_y^1 p_z^1$) has a larger first ionization energy, 1 400 kJ mol^{-1}, than carbon ($1s^2 2s^2 p_x^1 p_y^1$), 1 090 kJ mol^{-1}, since an electron is being removed from a similar orbital against a larger nuclear charge. It also has a larger first ionization energy than oxygen ($1s^2 2s^2 p_x^2 p_y^1 p_z^1$), 1 310 kJ mol^{-1} since the fact that the electron to be removed is paired more than offsets the effect of the increased charge of the oxygen nucleus.
9 D In all cases the second ionization energy must exceed the first. The ratio is largest for potassium of the elements listed since the second electron removed has a lower principal quantum number than the first.

$$\text{Ratio for potassium} = \frac{3\,070\,\text{kJ mol}^{-1}}{418\,\text{kJ mol}^{-1}}$$

10 A Cs$^+$ has the largest radius of the cations and F$^-$ the smallest radius of the anions listed.
11 E Descent of any major group in the periodic table shows an increase of atomic radius with atomic number. The lanthanides show a decrease. Some will define the lanthanides as the elements Cerium to Lutetium; use of lanthanum in the question may lead to less confusion and does not affect the answer.
12 B $^{222}_{86}\text{Ra} \xrightarrow{\alpha} {}^{218}_{84}\text{Po} \xrightarrow{\alpha} {}^{214}_{82}\text{Pb} \xrightarrow{\beta} {}^{214}_{83}\text{Bi}$
13 A The ground state of the carbon atom (i) has the lowest potential energy. (ii), (iii) and (iv) represent successive promotions of an electron.
14 A
15 B
16 A The properties affected will be those which are mass dependent and those which relate to atoms or molecules in close proximity to one another.

17 D
18 A The ionization energy of hydrogen is equivalent to that of a photon in the ultra-violet region.
19 B Argon
20 D Germanium
21 C Iron
22 B Argon (18 electrons)
23 B
24 E
25 E
26 C The capture of an electron from the 1s orbital (K shell) by the nucleus.
27 B The first electron affinities of the first row elements are lower than those of the second row elements. The difference between fluorine (-348 kJ mol^{-1}) and chlorine (-364 kJ mol^{-1}) is caused by the greater proximity of the added electron to other electrons in the case of the smaller atom.
28 A
29 B The second ionization energy of lithium, 7 300 kJ mol^{-1} is greater than that of helium, 5 250 kJ mol^{-1} since the electrons are being removed from the same orbital (1s) against a larger nuclear charge.
30 C In 1921 Stern and Gerlach split a beam of silver atoms and provided crucial evidence for the existence of unpaired electrons of opposite spin.

2 Structure and Bonding

1 D
2 B The (Pauling) electronegativity difference is greatest, 2.0, for KBr. For $MgCl_2$ the difference is 1.8. $MgCl_2$ hydrolyses in aqueous solution and this solution, on heating to 180°C, gives off HCl.
3 D The $HgCl_2$ molecule is linear with no lone pairs on the mercury atom.
4 C
5 A For the oxygen atom in water there are 2 lone pairs and 2 bonded pairs.
6 D
7 D In the vapour state the PF_5 molecules are trigonal bipyramidal.
8 E
9 C Silicon dioxide crystallises in a giant lattice.
10 D
11 B
12 E
13 A
14 D
15 C

16 A   The low electronegativity values make it certain that the compound is intermetallic. The scale used is that of Pauling.

17–21
For these questions the arrangement of *atoms*, not electron pairs, around the underlined atom is asked. The structures may be deduced by considering the numbers of bonded and non-bonded electron pairs.

17 D
18 A
19 C
20 E
21 B

22–25
The questions ask for the most useful model for interpretation of a specific property of the given substance.

22 C
23 D
24 E
25 B
26 C   *n*-propanol: molecular mass 60
                    boiling point 97°C
        ethanethiol: molecular mass 62
                    boiling point 36°C
27 C   pK_a for HF at 298 K = 3.25
        electronegativities (Pauling) F = 4.0, Cl = 3.0
28 D
29 A   Silicon tetrachloride is hydrolysed.
        Unlike carbon, silicon, having orbitals of principal quantum number 3, can accommodate more than 4 electron-pairs around the atom, as in SiF_6^{2-}
30 B   Thermal conductivity is not primarily a function of delocalised electrons.

3 The Gaseous State

1 E   There is a range of kinetic energies amongst the molecules of any gas sample at a temperature above 0 K.
2 D $0.75 \text{ g l}^{-1} \times 22.41 = 16.8 \text{ g}$
       The molecular mass of ammonia is 17.
3 B   At s.t.p. 0.65 g of the hydrocarbon will occupy a volume of 0.56 l (pressure doubled, temperature unchanged). 22.4 litres at s.t.p. will be occupied by

$$0.65 \times \frac{22.4}{0.56} = 26 \text{ g hydrocarbon}$$

4　C　Molecular mass $SO_2 = 64$

approximate density at s.t.p. $= \dfrac{64\,g}{22.41} \simeq 3\,g\,l^{-1}$

546 K and 2 atmospheres $\equiv$ 273 K and 1 atmosphere

5　C
$$\begin{array}{lc} & \text{He}:SO_2 \\ \text{molecular mass} & 4:64 \\ \sqrt{\text{molecular mass}} & 2:8 \end{array}$$

mean velocity $\propto \dfrac{1}{\text{molecular mass}}$

mean velocity $SO_2 = \dfrac{1200}{4} = 300\,m\,sec^{-1}$

6　E

7 & 8

Rate of diffusion $\propto \sqrt{\dfrac{1}{\text{molecular mass}}}$

$$\begin{array}{lc} & H_2:O_2 \\ \text{molecular mass} & 2:32 \\ \sqrt{\text{molecular mass}} & \sqrt{2}:\sqrt{32} \\ & 1:4 \end{array}$$

Relative rates of diffusion (in moles) 4:1
Relative rates of diffusion (by mass) $4 \times 2 : 1 \times 32$
$\qquad\qquad\qquad\qquad\qquad\qquad\qquad\quad 1 \qquad\; 4$

7　B
8　D
9　E　Since $V = \dfrac{nRT}{P}$, deviations from ideality which result in a lower value of V owing to intermolecular attraction will also result in a lower value for R. Ammonia deviates most from ideality amongst the given gases. For hydrogen the product PV is greater than would be expected for an ideal gas.
10　C　At all temperatures above 0 K the range of velocities is to be expected in accordance with the Boltzmann distribution.
11　D　The ratio of the specific heats of gases is related to their atomicity since it is dependent on the ways in which the molecular species store energy.
12　C　Since the water level in the tube is h mm above that in the container, the partial pressure of the hydrogen is $P - \left(p + \dfrac{h}{13.6}\right)$ mmHg.
13　D　There must always be an increase in potential energy when molecules are separated.
14　C　Where P and V are fixed it follows that nT is a constant and $T \propto \dfrac{1}{n}$

15	B	It is recognised than an approximate knowledge of the boiling point of the liquid may be necessary for purposes of experimental design.
16	B	(*ii*) is true only below the critical temperature.
17	E	For an ideal gas PV is independent of pressure. For real gases the variation is non-linear.
18	A	
19	C	If the pressure or partial pressure of the neon is unaltered the number of neon-neon collisions will not change.
20	B	
21	B	Lowering the external atmospheric pressure must initially increase the rate of evaporation.
22	A	(*iv*) is true of all gases

23–30

    Box P   Box Q
1 *mole helium*   2 *moles hydrogen*
    4 g   4 g

23	D	$P \propto n$. Ratio is 1:2
24	A	Ratio 1:1
25	C	There are twice as many particles in Box Q as in Box P. The average *distance* between molecules must be proportional to the cube root of volume containing them.
26	A	The gases are at the same temperature.
27	D	Total K.E. = Average K.E. × number of molecules Ratio 1:2
28	E	Diatomic molecules have 3 degrees of freedom.
29	B	The number of collisions is proportional both to the number of particles present and to their velocity. Ratio $1:2\sqrt{2}$
30	D	Assuming isotopic simplicity. Ratio 1:2

4 Energetics

1	E
2	C

The ionization energies of magnesium and aluminium (in kJ mol^{-1}) are:

	First	Second	Third	Fourth
Mg	736	1 450	7 740	10 500
Al	577	1 820	2 740	11 600

3 B
4 B

$$\begin{array}{ll} & \Delta H \\ 2C + 2O_2 \rightarrow 2CO_2 & -2a \\ 2CO_2 \rightarrow 2CO + O_2 & +b \end{array}$$

Adding $2C + O_2 = 2CO$ $-2a + b$

$$\Delta H_f CO(g) = \frac{-2a+b}{2} = \frac{b-2a}{2}$$

5 B
6 C

$$Ce^{4+}(aq) + e^- \rightarrow Ce^{3+}(aq) \qquad E^{\ominus} \; +1.61$$
$$Fe^{2+}(aq) \rightarrow Fe^{3+}(aq) + e^- \qquad -0.77$$

Adding $Ce^{4+}(aq) + Fe^{2+}(aq) \rightarrow Ce^{3+}(aq) + Fe^{3+}(aq) \qquad +0.84$

7 C Atomic radii (picometers)
Li 152
Na 186
K 231
Rb 244
Cs 262
Ionization energies, hydration energies and melting points all decrease from Li to Cs. Rb has the minimum boiling point of the alkali metal series.

8 D
$$\qquad\qquad\qquad\qquad\qquad\qquad \Delta H$$
$$K^+(g) + I^-(g) \rightarrow KI(s) \qquad -645 \, kJ\,mol^{-1}$$
$$KI(s) \rightarrow K^+(aq) + I^-(aq) \qquad +22 \, kJ\,mol^{-1}$$
Hydration Energy
$$K^+(g) + I^-(g) \rightarrow K^+(aq) + I^-(aq) \qquad -623 \, kJ\,mol^{-1}$$

9 D The breaking of 4 C—H bonds requires $1\,648 \, kJ\,mol^{-1}$
The breaking of 6 C—H bonds requires $2\,472 \, kJ\,mol^{-1}$
The breaking of 1 C—C bond requires $2\,810 - 2\,472 = 388 \, kJ\,mol^{-1}$

10 A
11 A
12 C pK_a HCl = 7.4   (298 K)
HBr = 9.5
HI = 10
ΔH_f° HCl $-92.3 \, kJ\,mol^{-1}$
HBr $-36.2 \, kJ\,mol^{-1}$
HI $+25.9 \, kJ\,mol^{-1}$

13 D
14 C There is no volume change in the reaction.
Thus $\Delta H = \Delta U$
15 C
16 A
17 B
18 A
19 D
20 E
21 D Standard heat of formation Al_2O_3 (corundum) $\simeq -1\,700 \, kJ\,mol^{-1}$
22 B Heat of neutralisation (e.g. for $HNO_3 + KOH$) $= -57.3 \, kJ\,mol^{-1}$
23 E Standard heat of formation for nitric oxide is $\simeq +90 \, kJ\,mol^{-1}$
24 A Hydrogen bond formation
25 A

26 C Heats of formation (kJ mol^{-1});
      $CaCl_2$ -795, $SrCl_2$ -828
      First ionization energies (kJ mol^{-1})
      Ca+590, Sr+549
27 C The first ionization energy of radium is $+510$ kJ mol^{-1}; only barium, $+502$ kJ mol^{-1} has a lower value amongst Group II elements.
28 D First ionization energies (kJ mol^{-1}); Na+494, Mg+736, Al+577.
29 A
30 E Cu_2O is well known and, unlike many cuprous salts, does not interact with water.

5 Kinetics

1 B
2 D The simple reaction $X + X \to X_2$ fulfils only the conditions of C and D.
     The sequence $X + X \to XY$ fast
     $XY + X \to X_2Y$ slow
     fulfils D but not C.
3 C
4 B For the rate determining step
     Rate $= k[VW]$
     but $[VW] \propto [V][W]$
5 A Since the reaction occurs more rapidly on the addition of Mn^{2+}(aq) the activation energy of the interaction of this species with one of the reactants must be lower than the activation energy of their reaction with one another. Mn^{2+} is in its lowest oxidation state in aqueous solution and thus cannot react with Tl^+(aq).
6 E The rate law cannot be deduced from the stoicheiometric equations.
7 C Initial rate $= k[P]^2Q$
     New rate $= k[2P]^2 \dfrac{Q}{2} = 2k(P)^2Q$
8 A Radioactive changes are first order.
9 E Photons (of blue but not of red light) atomise the chlorine molecules but not the hydrogen molecules.
10 E Halving the concentration of P will lower the rate by a factor of 4.
11 B The decreasing rate of interaction will be ion-ion, ion-molecule and ion-solid.
12 B
13 A
14 D
15 C
16 C
17 D For each photochemical reaction there is a threshold energy and photons of any energy equal to or greater than this may initiate the change. Both (*ii*) and (*iii*) are true of some photochemical reactions, but they cannot both be true of the same reaction.

18 A
19 C   The magnitude of the enthalpy change is the same for each reaction represented.
20 E
21 E   At all temperatures the reaction with the lowest activation energy must be the one which proceeds most rapidly (provided that any of the reactions take place to any extent).
22 A   At all temperatures the reaction with the highest activation energy must be the one which will proceed least rapidly.
23 E   The half-life of a radioactive substance does not vary over the stated temperature range.
24 A   Although the rate of formation of PCl_5 increases, its rate of dissociation increases more rapidly.
25 C   The reaction rate continues to rise until the enzymic catalyst can no longer function.
26 B
27 C
28 E   In a rigid container the addition of an inert gas will not affect the partial pressures of the reacting gases.
29 C
30 C ΔG is negative for this reaction ($-285\,\text{kJ mol}^{-1}$ at 298 K) but the high activation energy prevents reaction.

6 Equilibria I

1 C $\dfrac{[H^+][X^-]}{[HX]} = 1.6 \times 10^{-4}$

   $[HX] = 0.1$ and $[H^+] = [X^-]$
   Thus $[H^+]^2 = 1.6 \times 10^{-5} = 16 \times 10^{-6}$

2 D $\dfrac{[NH_4^-][OH^-]}{[NH_4OH]} = 10^{-5}$

   $[NH_4OH] = 1$ and $[NH_4^+] = [OH^-]$
   Thus $[OH^-]^2 = 10^{-5}$
   and $[OH^-] \simeq 3 \times 10^{-3}$
   giving pH $\simeq 11$
   In any solution in which the concentrations of ammonia and NH_4^+ are the same (as in A) $[OH^-] = 10^{-5}$, pH = 9

3 C
4 B

8

5   E $V + W \rightarrow X + Y$
         initially   1   1   0   0 moles
         at equilibrium $(1-x)(1-x)$ x x moles
         $$K = \frac{[X][Y]}{[V][W]} = \frac{x^2}{(1-x)^2} = 81$$
         $$\frac{x}{(1-x)} = 9$$

6   A $$\frac{[CH_3NH_3^+][OH^-]}{[CH_3NH_2]} = 49 \times 10^{-5}$$
         $[CH_3NH_2] = 0.1$ and $[CH_3NH_3^+] = [OH^-]$
         Thus $[OH^-]^2 = 49 \times 10^{-6}$

7   B
8   B
9   D $HY \rightarrow H^+ + Y^-$
         concentrations at equilibrium 10^{-2} 10^{-4} 10^{-4}
         $$K = \frac{(10^{-4})^2}{10^{-2}} = 10^{-6}$$

10 A   For phosphoric acid
         $K_1 = 1.1 \times 10^{-2}$
         $K_2 = 2.0 \times 10^{-7}$
         $K_3 = 3.6 \times 10^{-13}$
         The ratio K_1/K_2 is always large since it is more difficult to remove a proton from a negatively charged species than from a neutral one. Note that the ratio $K_2/K_3 \simeq K_1/K_2$

11 C $P + Q \rightarrow R + S$
         moles at equilibrium $\frac{1}{3}$ $\frac{1}{3}$ $\frac{2}{3}$ $\frac{2}{3}$
         $$K = \frac{(\frac{2}{3})^2}{(\frac{1}{3})^2} = 4$$

12 C
13 D
14 C
15 E   Henry's Law (an ideal gas law) is approximately obeyed only when the species in solution is the same as the species in the gas phase. It assumes no interaction between solute and solvent molecules.
16 A
17 A
18 B   For sulphuric acid pK_2 is 1.80.
19 A   Phosphoric acid (pK_1 2.15) is slightly weaker than phosphorous acid (pK_1 2.00). This may be deduced from a comparison of the structures knowing that in H_3PO_3 one of the hydrogen atoms is bonded directly to the phosphorus atom.
20 C
21 A H_3PO_4 is the only tribasic acid. With each of the other acids there will be an excess of sodium hydroxide solution.
22 C
23 A   Above 800°C the chloride exists as the monomer.

24	E	
25	D	
26	D	
27	B	pK values at 298 K CH_3COOH 4.76 $CH_2ClCOOH$ 2.86
28	C	$H^+(aq) + OH^-(aq) \rightarrow H_2O(l)$ $\Delta H = -57\,kJ\,mol^{-1}$ $H_2(g) + \frac{1}{2}O_2(g) \rightarrow H_2O(g)$ $\Delta H = -285\,kJ\,mol^{-1}$
29	B	The density of aniline at 293 K is 1.022 g cm^{-3}.
30	D	

7 Equilibria II

1	C	$PCl_5 \rightleftharpoons PCl_3 + Cl_2$ moles at equilibrium $(1-x)$ x x where x is the degree of dissociation $\dfrac{(1+x)}{1} = \dfrac{208}{156} = \dfrac{4}{3}$
2	C	$N_2O_4 \rightleftharpoons 2NO_2$ moles at equilibrium $(1-x)$ $2x$ where $x = 0.5$, 0.5 1 $\dfrac{1.5}{1} = \dfrac{92}{(\text{apparent M.W.})}$
3	E	
4	D	A given volume of 0.01 M acetic acid contains more solute particles than 0.01 M sucrose.
5	D	
6	A	
7	B	
8	C	
9	C	1 cm^3 excess M NaOH in approximately 100 cm^3 of solution.
10	C	For phosphoric acid $pK_1 = 2.15$, pK_2 7.21, pK_3 12.36 at 298 K
11	D	
12	C	$pK_a = -\log_{10}(K_a\,mol\,dm^{-3})$
13	A	
14	A	(iv) is not a sufficient condition.
15	A	
16	B	
17	D	
18	A	At $-0.40°C$ the concentration of the solution has increased by a factor of four.
19	D	
20	C	
21	D	$Cu^+(aq)$ disproportionates.
22	C	The blue colour is believed to be due to $CoCl_4^{2-}(aq)$
23	C	

24	E	$0.4\,g = 10^{-2}\,mol\,NaOH$. $pH = 12$
25	A	
26	C	
27	C	
28	B	
29	B	
30	D	

8 The Periodic Table

1	C	Electron configuration $1s^2 2s^3 p^9$
2	D	$X\ 1s^2\ Y\ 1s^3 2s^1$
		hence $XY(X^- Y^+)$
3	B	The second diatomic gas with a single bond would be element 14, the next element which requires only one electron to reach an inert gas configuration. Multiple bonding could not conceivably produce a diatomic gas of atomic number 10 or below containing $2p$ electrons.
4	D	The number of elements is a function of nuclear structure. Since the number of elements does not alter, the proportion of transition metals must change.
5	C	Data (picometers):

	Ionic Radii	Atomic Radii
Li^+	68	152
Be^{2+}	35	112
N^{3-}	171	75
O^{2-}	140	73
F^-	136	72

6	B	
7	E	
8	C	
9	E	The electronegativity difference between the first and last members of a group tends to increase with the group number. The values for the groups cited are IA 0.3, IIIA 0.2, IVB 0.7, VB 1.1, VIB (oxygen 3.5, polonium 2.0) 1.5. For VIIB the value is 1.8.
10	B	
11	B	
12	E	
13	A	
14	D	
15	C	
16	E	
17	A	
18	C	

19 E Solubilities g/100 g water, 293 K:
      BeSO$_4$ 39
      BaSO$_4$ 2.4×10^{-4}
      Mg(OH)$_2$ 9×10^{-4}
      Ba(OH)$_2$ 3.9
20 E
21 E
22 A
23 D } Note that no numerical scales have been attached to the axes on the diagrams.
24 C
25 B
26 B
27 B
28 C Many similarities are found, e.g. TlCl is "insoluble" in water and darkens on exposure to light.
29 B
30 C

9 The Chemistry of the Non-Metals

1 D
2 D H$_3$AsO$_3$ pK_1 9.22
     H$_3$AsO$_4$ pK_1 2,30, both at 298 K
3 E The sum of the first and second electron affinities of oxygen is $+702$ kJ mol^{-1} ($-142 + 844$).
     The lattice energy of, for example, magnesium oxide is 3 889 kJ mol^{-1}.
4 D (NH$_4$)$_2$CO$_3$(s) → 2NH$_3$(g) + CO$_2$(g) + 2H$_2$O(g)
5 C
6 C (NH$_4$)$_2$Cr$_2$O$_7$(s) → N$_2$(g) + 4H$_2$O(g) + Cr$_2$O$_3$(s)
7 A
8 A
9 B
10 A
11 D
12 D values in kJ mol^{-1}:

	F	Cl
electron affinities	-348	-364
first ionization energies	1 680	1 260
bond enthalpies at 298 K	158	242

13 A
14 A
15 A
16 B (ii) decrease, (iv) do not vary.
17 A
18 B The reduction of the length of the carbon chain in an organic compound.
19 A

20 E
21 B
22 E
23 A For carbonic acid pK_2 = 10.2 and NaOH is thus unsuitable as a titrant.
24 E
25 A
26 D
27 C
28 D
29 C
30 C

10 The Chemistry of the Metals

1 C Sodium and potassium do not react directly with gaseous nitrogen to form nitrides. These compounds are made by striking an electric arc between a platinum cathode and alkali metal anode under liquid nitrogen.

2 E The standard reduction potentials are:
$Li^+(aq)/Li$ -3.04 V
$Na^+(aq)/Na$ -2.71 V
$K^+(aq)/K$ -2.92 V

3 D Solid magnesium and calcium bicarbonates are unknown.
4 A $Rb_3N + 2NH_3 \rightarrow 3RbNH_2$
5 C
6 C
7 A The standard reduction potential for beryllium, $Be^{2+}(aq)/Be$, $E^\ominus -1.85$ V), is much less than for other members of the family.

8 C ⎧ In the formation of $CaCl_2(s)$ in preference to $CaCl_3(s)$ the crucial factor is the energy required to remove the third electron from the calcium atom. This is not compensated for in the larger lattice
9 D ⎨ energy of $CaCl_3$. The lattice energy of $CaCl_2$ is, however, sufficiently larger than that of CaCl(s) to offset the second ionization energy of calcium and thus CaCl(s) is not formed.

10 A Dilute sulphuric acid reacts slightly with deoxidised aluminium.
11 C In aqueous solution $3AuCl \rightarrow AuCl_3 + 2Au$
12 C
13 A
14 B Solubilities (g/100 g water at 293 K)
$Ba(NO_3)_2$ 8.7 $Ba(OH)_2$ 3.9
15 A
16 A Complexes formed: $Ag(CN)_2^-(aq)$
$Ag(NH_3)_2^+(aq)$
$Ag(S_2O_3)_2^{3-}(aq)$
17 A

18 E
19 E
20 B
21 C
22 D
23 B Boron hydrides are electron-deficient.
24 D Melting points: $SnCl_2$ 246°C, $SnCl_4$ −33°C
25 A Aluminium(III) chloride exists as Al_2Cl_6 from the temperature at which the solid sublimes (183°C) until it dissociates (c. 750–800°C) to give $AlCl_3$.
26 B Boron nitride is structurally similar to graphite with which it is iso-electronic. A diamond-like form, Borazon, claimed to be harder than diamond, has been produced.
27 A
28 D The $E^\ominus$ values would not support the validity of the reason as stated (see question 2). This is due to the large hydration energy of lithium which, as the metal, is the weakest reducing agent in Group IA.
29 C The electronegativities difference (1.5) corresponds (on the Pauling Scale) to 43% ionic character.
30 A

11 General Inorganic Chemistry

1 C 5.0 cm³ of 0.01M NaOH = 5×10^{-5} mol NaOH
Approximate molecular mass = $\dfrac{1 \times 10^{-2}}{5 \times 10^{-5}}$

2 D The sulphate ion is the only species listed containing bonds in which both electrons are supplied by one of the atoms bonded.

3 A Solubilities (g/100 g water, 293 K): CaF_2 0.0016, $CaCl_2$ 74.5.

4 C

5 B
$$H_2O_2 \rightarrow H_2O + \tfrac{1}{2}O_2$$
oxidation states of oxygen: −1 −2 0

6 C $Cu^{2+}(aq) + S^{2-}(aq) \rightarrow CuS(s)$

7 E Densities (g cm⁻³): Zn 7.1, Cd 8.6, Hg 13.6.

8 C

9 D In $Fe(CN)_6^{3-}$ the iron atom is one electron short of a noble gas configuration.

10 C
11 C
12 B Salts of strong bases and weak acids
13 A
14 E PH_3 is basic, SiH_4 does not interact with water.
15 A
16 A Potassium

17	D	Krypton
18	B	Vanadium
19	C	Bromine
20	B	
21	D	
22	E	
23	A	
24	C	pK_a methyl orange 3.7 (298 K)
25	B	Densities (g cm^{-3}): rhombic 2.07, monoclinic 1.96.
26	B	
27	C	
28	C	
29	D	Boiling points: NH$_3$ − 33.4°C PH$_3$ − 87.4°C
30	A	

12 Hydrocarbons

1	D	2C$_2$H$_6$ + 7O$_2$ = 4CO$_2$ + 6H$_2$O 20 cm^3 of ethane yield 40 cm^3 carbon dioxide and uses 70 cm^3 oxygen. 50 cm^3 of gas thus remain after cooling.
2	D	
3	B	CH$_3$CH$_2$CH═CH$_2$, CH$_3$CH═CHCH$_3$ (*cis-trans*) and (CH$_3$)$_2$C═CH$_2$
4	A	
5	B	
6	B	
7	C	*n*-pentane boils at 36.3°C, (CH$_3$)$_4$C at 9.5°C and (CH$_3$)$_2$CHCH$_2$CH$_3$ at 27.9°C (all at 1 atm.)
8	B	
9	D	
10	B	
11	A	
12	E	C$_6$H$_6$Br$_6$ is not an aromatic compound.
13	C	
14	D	the only ethyl ester.
15	E	Kolbe electrolysis
16	C	
17	E	
18	D	
19	D	
20	A	
21	D	209.1 kJ mol^{-1}
22	C	240.3 kJ mol^{-1}
23	E	120.15 kJ mol^{-1}. The "resonance energy" is A–D.
24	D	

25 D $C\equiv C$, 837 kJ mol^{-1}; $C=C$, 612 kJ mol^{-1}.
26 A
27 B
28 A
29 B
30 B   Naphthalene melts at 80.2°C.

13 Compounds of Carbon, Hydrogen and Oxygen

1 C   C boils at 82.2°C. The unsaturated alcohol (E) boils at 121°C (1 atm).
2 C   This allows 16 possible stereoisomers.
3 D $C=O$ in carbon dioxide = 122 pm
      $C—O$ in dimethyl ether = 143 pm
      $C\cdots O$ in the acetate ion = 130 pm
4 B pK_a values at 298 K are:
      propanoic   4.87
      acetic   4.76
      monochloroacetic   2.86
      ethanedioic (oxalic) 1.23 (pK_1)
      propanedioic (malonic) 2.83 (pK_1)
5 B
6 D
7 C   Oxidation of ketones involves the breaking of C—C bonds and, with the exception of the haloform reaction, takes place only under severe conditions.
8 A
9 E
10 D
11 A No conditions are specified and each of the compounds, except diethylketone, is likely to be present in some degree.
12 B The pK_a values (298 K) are as follows:
      phenol   10.00
      o-nitrophenol   7.21
      o-cresol   10.28
      p-cresol   10.26
      2,4,6-trinitrophenol 0.42
13 C
14 A
15 A
16 B
17 B
18 E
19 A
20 B Crotonaldehyde (buten-2-al) is produced by heating the product of the aldol condensation.

21 E
22 D
23 A The boiling points are $-24.8°C$ and $78.5°C$ (1 atm.).
24 E Propanoic acid boils at $141°C$, methyl ethanoate at $57.3°C$ (1 atm.).
25 C
26 A The CN^- acts as a nucleophile and a highly acidic medium retards the reaction. In practice the reaction is often carried out using sodium cyanide and mineral acid.
27 A
28 A The bromide ion is the nucleophile and protonated ethanol is the substrate.
29 B
30 B

14 The Organic Chemistry of Nitrogen and the Halogens

1 C
2 D
3 C Acidity and basicity constants are low for amino acids in comparison with the unsubstituted acid. This is reasonable when it is realised that the K_a value refers to a substituted ammonium ion RNH_3^+.
4 D C_2H_5Cl, $C_2H_4Cl_2(2)$, $C_2H_3Cl_3(2)$ $C_2H_2Cl_4(2)$, C_2HCl_5, C_2Cl_6
5 D
6 A There are only α and β positions in a monosubstituted naphthalene.
7 D
8 C
9 A
10 A
11 C (*i*) and (*iii*) exhibit geometrical isomerism.
12 E The replacement of the diazonium group by Cl or Br requires the use of a Cu(I) salt.
13 E
14 B
15 C
16 D Like the preceding question, this is a Friedel-Crafts alkylation.
17 E Secondary amines, both aromatic and aliphatic, yield *N*-nitrosoamines with nitrous acid. In this case *N*-nitroso-*N*-methylaniline is formed.
18 B The only primary aromatic amine.
19 C
20 A The primary aliphatic amine yields a complex mixture of products with nitrous acid.
21 B

22 A
23 D
24 E
25 C
26 C
27 C The nitrogen is originally present in the amine e.g. hexamethylene diamine → nylon 6:6.
28 C Formyl chloride exists only at very low temperatures.
29 E
30 E K_b (aniline) $= 4.2 \times 10^{-10}$
      (methylamine) $= 4.4 \times 10^{-4}$
      both at 298 K.

15 General Organic Chemistry

1 C $C_3H_8 + 5O_2 = 3CO_2 + 4H_2O$
2 C
3 A A boils at 27.9°C (1 atm.); n-pentane (C) boils at 36.3°C.
4 E The triphenyl radical is the classic example of a "stable" free radical.
5 D
6 B
7 C
8 A
9 E C:H $= 7:14$ or $1:2$, hence CH_2.
10 A
11 D
12 A A carbonium ion may be defined as a group of atoms that contains a carbon atom bearing only six electrons.
13 B
14 C Aryl halides differ from alkyl halides in exhibiting a low reactivity towards nucleophilic reagents. The presence of certain groups in the benzene ring can activate the halogen atom towards replacement. At 298 K, pK_a (benzoic) $= 4.20$ and pK_a (p-nitrobenzoic acid) $= 3.43$.
15 B The K_b values (298 K) are:
      ammonia 1.8×10^{-5}
      methylamine 4.4×10^{-4}
      dimethylamine 5.1×10^{-4}
      aniline 4.2×10^{-10}
      diphenylamine 0.7×10^{-13}
16 D
17 D
18 E
19 E
20 D
21 C

22 B
23 E
24 D
25 C
26 A
27 A
28 A
29 C
30 A

16 Experimental Procedures

1 D
2 E
3 D
4 C The least precise measurement will be the burette reading and this will therefore govern the accuracy of the final result.
5 D Acetamide melts at 82.3°C (1 atm.).
6 E Silver fluoride has a solubility of 195 g per 100 g water at 18°C.
7 D This will produce two "equivalents" of nitric acid.
8 C
9 E
10 E Percentage uncertainty is approximately 4%. In D it is less than 1%.
11 C
12 E The permanganate is assumed to be in the burette.
13 C
14 A
15 B
16 D Ethanol/water is a common mixed solvent.
17 D
18 D
19 C ZnS is not precipitated in acid solution.
20 B
21 A Zinc hydroxide dissolves in excess aqueous ammonia.
22 D Mg^{2+} is not precipitated under these conditions.
23 A
24 B
25 C
26 E
27 C NaOH is more corrosive than HCl. Neither should be pipetted by mouth.
28 A
29 A
30 D

17 Revision Paper I

1. D Values in kJ mol^{-1} are:
   H—H 435
   Cl—Cl 242
   Br—Br 193
   I—I 152
   C—C 348
2. E
3. B
4. A $H^+(aq) + OH^-(aq) = H_2O$
   $4OH^-(aq) \equiv 2H_2O + O_2 + 4e^-$
5. E
6. D Addition of $SnCl_2$ to $HgCl_2$ solution precipitates white mercury(I) chloride. If the tin salt is in excess, mercury may be formed.
7. A
8. D
9. B
10. C
11. C
12. D
13. D
14. D
15. D
16. E
17. D
18. A
19. A pK_a for $Al(H_2O)_6^{3+}$ is 5.0 at 298 K.
20. D The dipole moment of SO_2 is 1.61 Debye units.
21. A $Cl_2(g) + 2I^-(aq) \rightarrow 2Cl^-(aq) + I_2(s)$
22. A
23. B
24. A
25. D
26. B
27. A The boiling points are:
   H_2Se $-41.7°C$
   H_2Te $-1.8°C$
28. B The m.p. of graphite is quoted as 4 003 K.
29. E
30. D Methylamine is a gas at room temperature
   b.p. = $-6°C$ (1 atm).

18 Revision Paper II

1	C	Benzenecarbaldehyde oxime has *cis* and *trans* forms, melting at 35° and 125°C respectively.
2	B	A Cannizzaro reaction, given by aldehydes containing no α hydrogen atoms.
3	E	$CH_3COONH_4 \rightarrow CH_3CONH_2 + H_2O$. The salt is usually heated under reflux with glacial acetic acid.
4	A	A diazonium salt is first formed.
5	E	For example $CH_3CHBrCH_3$ and $CH_3CH_2CH_2Br$
6	B	
7	C	$(CH_3)_3 \overset{+}{N}HCl^-$
8	A	CO_2 is a linear molecule.
9	C	
10	E	
11	B	
12	D	
13	A	
14	D	
15	D	$3OCl^-(aq) \rightarrow 2Cl^-(aq) + ClO_3^-(aq)$
16	C	
17	D	
18	D	Note that Le Chatelier's principle applies only to an equilibrium system, *i.e.* to the saturated solution with excess solute present.
19	B	$NaH + H_2O = NaOH + H_2$. The amide yields NH_3.
20	A	
21	D	Note in (*i*) that the spectrum is not that of the argon but of the incandescent filament.
22	A	
23	B	
24	E	
25	C	
26	A	
27	B	$PbSO_4 = 0.004$ g; $CaSO_4 = 0.21$ g; $LiSO_4 = 35$ g; $Al_2(SO_4)_3 = 36.2$ g/100 g water, 293 K.
28	A	
29	C	
30	D	